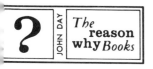

JOHN DAY | The reason why Books

INTEGERS
POSITIVE AND NEGATIVE

Irving Adler

Illustrated by Laurie Jo Lambie

The John Day Company • *New York*

THE "REASON WHY" BOOKS
AIR
ATOMIC ENERGY
ATOMS AND MOLECULES
THE CALENDAR
COAL
COMMUNICATION
DIRECTIONS AND ANGLES
ENERGY
THE EARTH'S CRUST
EVOLUTION
FIBERS
HEAT
HOUSES
INSECTS AND PLANTS
INTEGERS: POSITIVE AND NEGATIVE
IRRIGATION: CHANGING DESERTS TO GARDENS
LANGUAGE AND MAN
LEARNING ABOUT STEEL THROUGH THE STORY OF A NAIL
MACHINES
MAGNETS
NUMBERS OLD AND NEW
NUMERALS: NEW DRESSES FOR OLD NUMBERS
OCEANS
RIVERS
SETS
SHADOWS
STORMS
TASTE, TOUCH AND SMELL
THINGS THAT SPIN
TREE PRODUCTS
WHY? A BOOK OF REASONS
WHY AND HOW? A SECOND BOOK OF REASONS
YOUR EARS
YOUR EYES

Library of Congress Cataloging in Publication Data

Adler, Irving.
    Integers: positive and negative.

    (His The Reason why books)
    SUMMARY: Explains the meaning of integers and discusses their uses. Includes directions for making a slide rule and a simple adding machine.
    1. Numbers, Natural—Juvenile literature.
[1. Numbers, Theory of]   I. Lambie, Laurie Jo, illus. II. Title.
PZ10.A31o   512'.72   78–162589

The John Day Company, 257 Park Avenue South, New York, N.Y. 10010
An Intext Publisher
Published on the same day in Canada by Longman Canada Limited.
Printed in the United States of America

U. S. 1719344

Contents

Rewards and Penalties

There are some games, such as *ring toss*, in which at each play you can either win some points or win no points at all. The points you win are a *reward*, and are represented by the counting number that tells how many you have won. If you win no points at all there is no reward, and the absence of a reward is represented by the number zero. To find your score you add the winnings of all your plays.

There is another kind of game in which there is another possible outcome besides winning points or winning no points at all. You might receive a *penalty*, which is a number of points to be taken away from the points you have already won or the points you may win in the future. In a game like this, while a counting number is used to represent a *reward*, another kind of number is needed to represent a *penalty*. The purpose of this book is to introduce you to this new kind of number, and to show you how the old and new numbers may be added to keep score. It will also show some important uses of these new numbers besides scoring games with rewards and penalties.

4

A Coin Toss Game

Two people play a game in this way: One player tosses a coin five times. After each toss, the coin is covered while the other player guesses whether it is *heads* or *tails*. The guesser wins a reward of one point for each guess that is right, and a penalty of one point for each guess that is wrong. Then it is his turn to toss while the other player guesses. To figure the winnings of a player at each play, the rewards and penalties are combined according to this rule: A penalty of one point and a reward of one point cancel each other, or wipe each other out.

Here is a way of using checkers to figure the score for a play. Use a black checker to represent a reward of one point, and use a red checker to represent a penalty of one point. Suppose a player guessed right two times and guessed wrong three times. Then put out two black checkers to show his two point reward, and put out three red checkers to show his three point penalty. Each black checker cancels one red checker. To show that they have been cancelled, remove them from the table. Continue

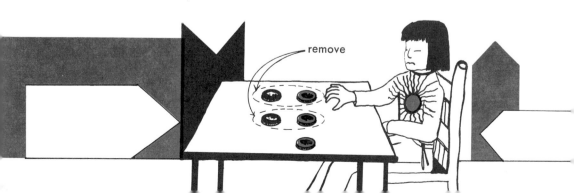

to remove cancelled checkers, one black and one red each time, until all the checkers of one color have been removed. The checkers that remain show the score for the play. In this case one red checker is left, so the score is a penalty of one point.

The drawings below show some other possible outcomes of one play when a coin is tossed five times. Use checkers to show these outcomes, or copy the drawing on a separate sheet of paper, and then figure the score for each play. (Do not write in this book.)

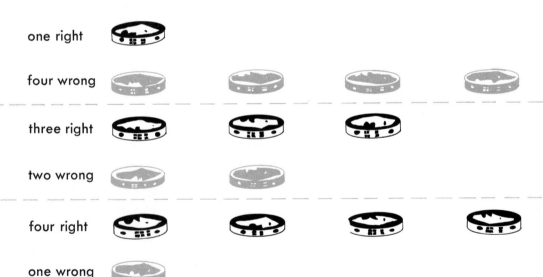

The game can also be played by tossing a coin *six* times for each play. Some possible outcomes of a six-toss play are shown on the next page. Use checkers to figure the score for each play.

6

one right

five wrong

two right

four wrong

three right

three wrong

four right

two wrong

five right

one wrong

7

It is convenient to have written symbols (SIM-buhls) that stand for rewards and penalties, so that we may keep score in writing as we play a game.

The numerals for *counting numbers* can be used as symbols for *rewards*. We can use the numeral 1 to stand for a reward of one point, the numeral 2 to stand for a reward of two points, the numeral 3 to stand for a reward of three points, and so on.

To stand for a score that is a penalty, we use a symbol that is made up of two parts. One part is the numeral that tells us how many points are in the penalty. The other part is a sign that tells us that the score is a penalty, and not a reward. A convenient sign to use for this purpose is a *minus sign*, to remind us that a penalty *takes away* some of the points you have won or will win as rewards. The two parts are put together in this way: The minus sign is written to the left of the numeral, and near the top of the numeral. Thus, the symbol ⁻3 stands for a penalty of three points. We read this symbol as *negative three*. *Negative three* is a new kind of number, different from the counting number *three*, because *negative three* stands for a penalty of three points, while *three* stands for a reward of three points.

6

⁻4

How do you read each of these symbols, and what does it stand for?

(a) ⁻5   (b) ⁻2   (c) ⁻7

The new numbers, negative one, negative two, negative three, and so on, are called *negative integers* (IN-tuh-juhrs). The old numbers, one, two, three, and so on, we already know as counting numbers. But they are also called *positive integers* to remind us that they are different from the negative integers. A negative integer stands for a penalty of one or more points. A positive integer stands for a reward of one or more points. When the score is no points at all, neither a penalty nor a reward, we use the symbol 0 (zero) for it.

The positive integers, the negative integers, and zero are united in one big family of numbers called *the integers.*

In the drawing above, we show two ways of representing a reward and a penalty. One way uses checkers. The other way uses an integer.

In each drawing below, a reward or a penalty is represented by a row of checkers. What number stands for the same reward or penalty?

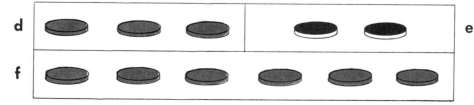

d .... e

f

Keeping Score with Integers

Two boys played a series of three games in which a coin was tossed several times. In the first game, each had his turn guessing heads or tails for five tosses. In the second game, each guessed for six tosses. In the third game, each guessed for seven tosses. They combined the scores for the three games to get a total score for the series. The score of each player for each game is shown below in two ways, by checkers and by integers. Put red and black checkers on a table to show each score, and then remove the pairs of checkers that cancel. Count the checkers that are left to see if the integer shown in the table is correct.

To find the total score for the first player, put down checkers to represent the score for each game. What color will you use to represent −1? What color will you use to represent 2 and 1? Now remove pairs of checkers that cancel, one red and one black together each time. Continue removing checkers that cancel until all the checkers that are left have the same color. What color is it? How many checkers are left? What integer stands for the total? Find the total score of the second player too.

	first player				second player					
first game	● ○	● ○	● ○		−1	● ○	● ○	●			1
second game	● ○	● ○	●	●	2	● ○	● ○	● ○			0
third game	● ○	● ○	● ○	●	1	● ○	● ○	○	○	○	−3
			total score						total score		

Which Score Is Higher?

In the series of games described on page 10, the first player got a total score of 2, while the second player got a total score of ⁻2. Who won the series? The first player's final score is a reward of two points, and the second player's final score is a penalty of two points. We say that the first player won, because we think of a reward as being better than a penalty. So we say that the number 2 is greater than the number ⁻2.

Suppose that in another series the first player had a final score of ⁻3, and the second player had a final score of ⁻2. Who won that series? To help us answer this question, let us imagine that both players were given five free points before they played the series, and that their scores were to be combined with these five free points. In the drawing below, checkers are used to show the free points and the scores of the series. For the first player, his penalty of three points removes three of his free points, and leaves him with two points. For the second player, his penalty of two points removes two of his free points, and leaves him with three points. We see that a score of ⁻2 is better than a score of ⁻3, because it removes fewer points. So a score of ⁻2 wins over a score of ⁻3. For this reason we say that the number ⁻2 is greater than the number ⁻3.

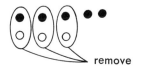

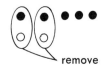

Free	5	● ● ● ● ●
Score	−4	◯ ◯ ◯ ◯

Free	5	● ● ● ● ●
Score	−5	◯ ◯ ◯ ◯ ◯

Free	5	● ● ● ● ●
Score	0	

Free	5	● ● ● ● ●
Score	3	● ● ●

Each drawing above shows a reward or a penalty being combined with five free points. Use checkers to show the same combination. Remove the checkers that cancel, and compare the sets of checkers that are left. Which score is higher, −4 or −5? −4 or 0? 0 or 3?

Use checkers to help you figure out which of the three scores 2, −2 and 0 is the lowest, and which is the highest. Write down the integers 2, −2, 0 in order, from the least to the greatest.

What is the next greater integer above 7? What is the next greater integer above −7?

Four More Games

Rolling a Die. This game is played with a one-inch wooden cube that is first prepared in this way: Paste a small square of masking tape on each face of the cube. Write the number 1 on one square and write −1 on the opposite square. Write 2 on one of the squares that is still empty, and write −2 on the opposite square. Write 3 on an empty square, and then write −3 on the last empty square. Two or more people play the game. First one player rolls the die, and then each of the other players rolls. The score for a roll is the number that is on the top face when the die stops rolling. Each player rolls five times in a game. To keep score, each player makes a chart like the one shown below:

	SCORE BEFORE THIS PLAY	SCORE OF THIS PLAY	TOTAL SCORE AFTER THIS PLAY
FIRST PLAY	0		
SECOND PLAY			
THIRD PLAY			
FOURTH PLAY			
FIFTH PLAY			

The score before the first play is 0, so 0 is written in the first box on the first line. After each play, the score is entered in the second box for that play. Checkers are used to combine that score with the score before that play. This total is written in the third box on the same line, and is written again in the first box on the next line. The total score after the fifth play is the player's score for the game. The player who ends with the highest score wins.

The Red and the Black. This game is played with ten playing cards: clubs 1, 2, 3, 4 and 5; and hearts 1, 2, 3, 4 and 5. There are two players. One player shuffles the

13

cards and then deals them out face down, one at a time, to each player in turn. Then each player turns up his top card, and removes it from his pile of cards. His score for the play is a reward if the card is black, and it is a penalty if the card is red. The number of points in the reward or penalty is the number on the card. For the next play, he turns up the next card, and so on. Each player writes his scores in a chart like the one on page 13, and figures his totals with the help of checkers. This game can also be played by four people, if all four suits— clubs, hearts, spades and diamonds—are used.

Coin Toss Solitaire. This is a game for one person playing by himself. For each play, shake three or more coins in a cup and toss them. Each head that comes up gives you a reward of one point. Each tail gives you a penalty of one point. Figure your score for the play by first removing pairs of coins whose points cancel (one head and one tail). Make five plays, and keep score in a chart like the one on page 13.

Checkers in a Bag. Put equal numbers of red and black checkers in a bag, and mix them. Any number of people may play this game. Each player in turn draws three checkers from the bag. He puts aside any checkers that cancel (one red and one black). What is left tells him the score for his play. Each player draws five times in turn, and keeps score in a chart.

14

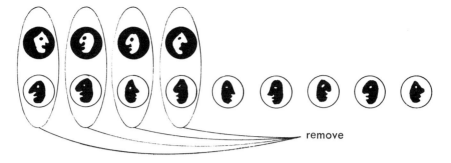

remove

Adding Integers by Using Checkers

All addition of integers can be done with the help of checkers. For example, to add 4 and $^-9$, put out four black checkers and nine red checkers, and then remove pairs of checkers that cancel each other (one red and one black), until checkers of only one color are left. In this case, as shown above, five red checkers are left, so $4 + (^-9) = ^-5$.

Use checkers to do the addition exercises pictured below.

| $(^-6) + 5$ | $(^-3) + 5$ | $4 + (^-4)$ | $(^-2) + (^-3)$ |

○ ○ ○ ○ ○   ○ ○ ○   ● ● ● ●   ○ ○

● ● ● ● ●   ● ● ● ● ●   ○ ○ ○ ○   ○ ○ ○

Use checkers to do these addition exercises:

$$8 + (^-9) \qquad (^-7) + 1 \qquad (^-8) + 1$$
$$0 + (^-4) \qquad (^-6) + 6 \qquad (^-3) + 0$$

Use checkers to do these pairs of addition exercises, and compare each pair of answers: $2 + 5$ and $5 + 2$; $3 + (^-5)$ and $(^-5) + 3$; $(^-2) + (^-4)$ and $(^-4) + (^-2)$. When two integers are added, does it make any difference which integer is written first?

15

Picturing Integers as Arrows

So far we have pictured an integer as a set of checkers —black checkers if the integer is positive, and red checkers if the integer is negative. It is sometimes useful to picture an integer in another way, as an arrow. To learn this other way, first prepare some cardboard arrows as follows: Cut strips of thin cardboard, each three fourths of an inch wide. Then cut each strip into pieces that are two inches long. At one end of each piece, cut the corners off to form the pointed *head* of the arrow.

tail  **head**

The end that is not pointed will be called the *tail* of the arrow. Using crayons, color one side of each arrow black, and color the other side red. We will think of the length of each of these arrows as *one unit*.

We can use these arrows to picture integers in this way: An arrow placed on the table with its black side up, and with its head pointing to the right, stands for the positive integer 1. An arrow with its red side up, and with its head pointing to the left, stands for the negative integer −1. The length of each arrow is one unit.

−1  **1**

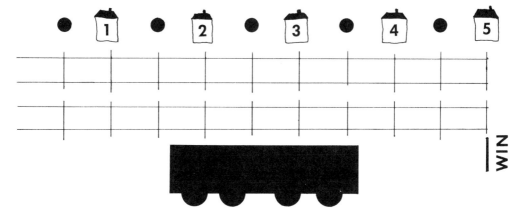

each space between stations between the lines marked *start* and *win*. Put a red checker in each space between stations between the lines marked *start* and *lose*. The name of a station shows how many checkers a train must pass to go to that station from the starting line. The name is a positive integer if the checkers are black, and the name is a negative integer if the checkers are red.

Each player begins with his car at the starting line on his track. The spaces between stations are thought of as *units* of distance. Each player in turn rolls the die. The number that turns up tells him how to move his car. If the number is a positive integer, it tells him how many units to move his car *forward* toward the winning line. If the number is a negative integer, it tells him how many units to move his car *backward* toward the losing line. If, after several moves, a car reaches or crosses the losing line, it is out of the race. A car wins the race if it is the first car to reach or cross the winning line, or if it is still in the race while all the other cars are out of it.

At each move in the game, say the integer that describes your move, name the station your car starts from, and name the station that it moves to. For example, if your car is at ⁻1 and you roll a 2, you say, "Two moves my car from negative one to one."

Answer these questions about the train race. To find the answers, move your finger along the track on pages 20 and 21.

1. If a car starts at station 1,
    what station will a move of 1 take it to?
    what station will a move of ⁻1 take it to?
    what station will a move of 2 take it to?
    what station will a move of ⁻2 take it to?
    what station will a move of 0 take it to?

2. If a car starts at station 3,
    what station will a move of 1 take it to?
    what station will a move of ⁻1 take it to?
    what station will a move of 2 take it to?
    what station will a move of ⁻2 take it to?
    what station will a move of 0 take it to?

3. If a car starts at station ⁻2,
    what station will a move of 1 take it to?
    what station will a move of ⁻1 take it to?
    what station will a move of 2 take it to?
    what station will a move of ⁻2 take it to?
    what station will a move of 0 take it to?

The Thermometer

We measure temperature with a thermometer. In the thermometer pictured on this page there is a liquid in a sealed glass tube. The liquid rises in the tube when it gets warmer, and it falls when it gets colder. Numbered lines are arranged alongside the tube like stations along a railroad track. The number at the level of the top of the liquid shows the number of degrees in the temperature. Positive integers are used to show temperatures that are warmer than or *above zero*. Negative integers are used to show temperatures that are colder than or *below* zero.

Move your finger along the scale of the thermometer to show a rise in temperature of 10 degrees; to show a rise of 20 degrees; to show a fall of 10 degrees; to show a fall of 20 degrees.

If the temperature rises 20 degrees from ⁻10 degrees, what does it end at? If the temperature drops 20 degrees from ⁻10 degrees, what does it end at?

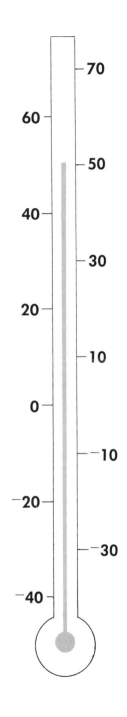

The Number Line

It is often useful to picture the integers as points on a line, arranged along the line the way stations in the train-race game are arranged along the track. A line that has the integers pictured on it in this way is called a *number line*. To make one, follow these directions: Draw a straight line that runs from left to right. Put a dot on the line and label it 0. Choose any unit of length, and put a dot on the line *to the right* of 0 so that its distance from 0 is one unit. Label this dot 1. Put another dot to the right of 1, so that its distance from 1 is one unit. The distance to this dot from 0 is two units, so label this dot 2. Continue in this way, putting each new dot one unit to the right of the last one you put down. The next labels will be 3, 4, 5, and so on, in that order. In each case the label of a dot tells the number of units in the distance from 0 to that dot. In this way we picture every *positive* integer as a dot or point that is *to the right* of 0 on the number line.

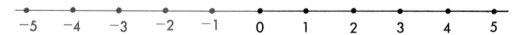

$$-5 \quad -4 \quad -3 \quad -2 \quad -1 \quad 0 \quad 1 \quad 2 \quad 3 \quad 4 \quad 5$$

Now start at 0 again, and put dots to the left of zero, so that each new dot is one unit to the left of the last one put down. Label these dots ‾1, ‾2, ‾3, and so on, in that order. In each case the minus sign in the label tells you that the dot is to the left of zero, and the number in the label, read without the minus sign, tells you how many units there are in the distance from 0 to that dot. In this way we picture every *negative* integer as a point that is *to the left* of 0 on the number line.

The number of units in the distance from 0 to any point is called the *absolute value* of the integer that is pictured by that point. Thus, the absolute value of 3 and the absolute value of ‾3 are both 3, because both have a distance of 3 units from the zero point. What is the

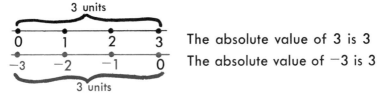

The absolute value of 3 is 3
The absolute value of ‾3 is 3

absolute value of 2? of ‾4? of 0? How many points on the number line have an absolute value equal to 6? How many points have an absolute value equal to 0?

One advantage of the number line is that we can see at a glance which of two integers is the greater: In any pair of integers, the integer that is on the right on the number line is the greater. For example, on the number line, 5 is to the right of 2, so 5 is greater than 2; 0 is to the right of ‾1, so 0 is greater than ‾1; ‾2 is to the right of ‾5, so ‾2 is greater than ‾5.

Which is greater, ‾9 or ‾6? 2 or ‾5? 0 or 7? 0 or ‾2?

Addition on the Number Line

Another advantage of the number line is that we can use it to do addition.

Two-step addition. To add two integers, use them as directions for moving on the number line in two steps, starting at 0. A positive integer means move to the right, and a negative integer means move to the left. The absolute value of an integer tells you how many units to move for that step. After you take both steps, the number of the point you end at is the sum of the two integers. For example, $2 + (^-3)$ means start at zero, move two units to the right, and then three units to the left. You end at the point labeled $^-1$, so $2 + (^-3) = ^-1$.

Two-step addition for $2 + (^-3)$

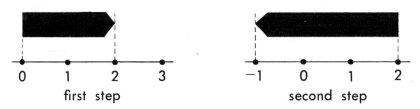

first step   second step

Find each of these sums by taking two steps on the number line, starting at zero:

$$3 + (^-2) \qquad 2 + 5 \qquad (^-2) + (^-2)$$
$$4 + 0 \qquad 0 + (^-2) \qquad (^-3) + (^-3)$$

Do each of these addition exercises in three different ways: (1) using red and black checkers; (2) using

arrows; and (3) by two steps on the number line:
$$(^-4) + 2 \qquad (^-3) + (^-4) \qquad 3 + (^-5)$$

In two-step addition, the first step always takes you to the point whose label is the first integer of the two that are being added. Knowing this, we can skip the first step, and start right off at that point. Then we can do the addition in one step in this way:

One-step addition for 5 + (−3) = 2

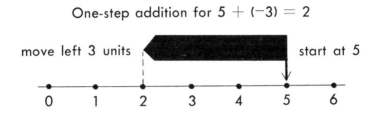

move left 3 units   start at 5

One-step addition. To add two integers, start at the point whose label is the first integer. Then move away from that point according to the directions given you by the second integer. The number of the point you end at is the sum of the two integers. For example, $5 + (^-3)$ means start at 5 and move 3 units to the left. You end at 2, so $5 + (^-3) = 2$.

Find these sums by one-step addition on the number line:
$$2 + (^-6) \qquad (^-4) + 4 \qquad 3 + (^-1)$$
What addition facts are pictured by these drawings of one-step addition?

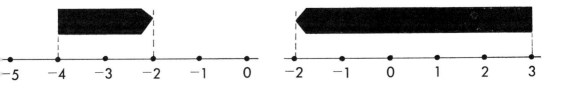

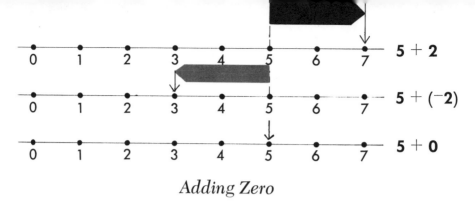

Adding Zero

Let's do these three addition exercises on the number line, and then compare them:

$$5 + 2 \qquad 5 + (^-2) \qquad 5 + 0$$

If we use one-step addition, we start at 5 in each case. Since 2 is a *positive integer* with absolute value 2, we find $5 + 2$ by moving two units *to the right* of 5. Since $^-2$ is a *negative integer* with absolute value 2, we find $5 + (^-2)$ by moving two units *to the left* of 5. Since 0 has absolute value 0, and is neither positive nor negative, we find $5 + 0$ by moving *no units at all, moving neither to the right nor to the left*. Then we stay at the 5, and the answer is 5.

To find $(^-2) + 0$, we start at $^-2$, and move no units at all, moving neither to the right nor to the left. Then we stay at $^-2$.

To add 0 to any integer, we start at that integer, and move no units at all, moving neither to the right nor to the left. Then we stay at that integer. So, *when 0 is added to an integer, the sum is that integer.*

Now let's examine sums like $0 + 2$ and $0 + (^-2)$.

Using one-step addition, to find $0 + 2$, we start at 0 and move 2 units to the right. Then we end at 2. To find $0 + (^-2)$, we start at 0 and move 2 units to the left. Then we end at $^-2$. We see that *if any integer is added to 0, the sum is that integer.*

Opposites

On the number line, the distance of 2 from 0 is 2 units. There is another number on the opposite side of 0 that is also 2 units from 0. It is $^-2$. We call $^-2$ the opposite of 2, and we call 2 the opposite of $^-2$. In general, if two integers have the same absolute value different from 0, the integers lie on opposite sides of 0, and we call each of them the opposite of the other.

What integer is the opposite of 3? What integer is the opposite of $^-7$? What integer is the opposite of $^-16$?

The number 0 is an exceptional number. Its absolute value, or distance from 0, is 0. It is the only integer whose absolute value is 0. So we say that 0 is its own opposite.

Find the following sums:

$$0 + 0 \qquad 1 + (^-1) \qquad 2 + (^-2)$$
$$(^-6) + 6 \qquad (^-3) + 3 \qquad (^-4) + 4$$

If you add any integer to its opposite, what is the sum?

29

Electrical Charges in the Atom

Integers are useful for showing and combining the electrical charges in an atom.

All material things are made of atoms.* There are three different kinds of particles in an atom, *protons*, *neutrons*, and *electrons*. The protons and neutrons are all crowded together in the central core of the atom, called its *nucleus*. The electrons revolve around the nucleus, the way the earth revolves around the sun.

There are two different kinds of electrical charges. One kind is called *positive*, and the other kind is called *negative*. Every proton has a positive electrical charge. Every electron has a negative electrical charge. Every neutron has no electrical charge at all, so we say it is electrically neutral.

If we use the charge on a single proton as the unit for measuring charge, then the charge on every proton is 1, the charge on every electron is $^{-}1$, and the charge on every neutron is 0. The total charge on 2 protons is 2. The total charge on 2 electrons is $^{-}2$.

What is the total charge on 3 protons? on 4 protons? on 3 electrons? on 4 electrons? on 3 neutrons? on 4 neutrons?

* See *Atoms and Molecules*, by Irving and Ruth Adler, The John Day Company, New York.

The nucleus of an ordinary helium atom contains 2 protons and 2 neutrons. What is the total charge on the nucleus?

The nucleus of an ordinary sulfur atom contains 16 protons and 16 neutrons. What is the total charge on this nucleus?

An atom of ordinary hydrogen has 1 proton in its nucleus, and 1 electron revolving around the nucleus. The charge on the nucleus is 1, and the charge on the electron is -1. The total charge on the atom is $1 + (-1)$. What integer stands for this charge?

An atom of ordinary carbon has 6 protons and 6 neutrons in its nucleus, and 6 electrons revolving around the nucleus. What is the total charge of the nucleus? What is the total charge of the electrons? What is the total charge of the whole atom?

An atom that has as many electrons as there are protons in its nucleus has a total charge of 0, so it is electrically neutral. If a neutral atom loses some of its electrons, or gains some extra ones, it then has a charge that is not 0, and it is called an *ion* (EYE-on).

Each drawing below shows the total charge on the nucleus of an atom, and the total charge of the electrons that surround the nucleus. Find the total charge of the whole atom in each case. Which of them are neutral atoms? Which of them are ions?

Charges on atoms and ions

Money Received and Money Spent

Integers are useful for keeping records of money received and money spent. *Positive integers can be used to stand for money received, and negative integers can be used to stand for money spent.*

For example, if you earned $5 and spent $3 during a certain week, you could use 5 to stand for the number of dollars you earned, and ⁻3 to stand for the number of dollars you spent. If you add these two numbers, the sum would show you the number of dollars you would have left after you paid your bills:

$$5 + (^-3) = 2.$$

If you had earned $3, and spent $5, you wouldn't be so well off. The $3 you earned would pay for only part of what you had bought. After you paid the $3, there would still be $2 that you owe. Instead of having money left over, you would have a debt. This fact is shown very easily if you use integers to stand for what you earned and what you spent. If you let 3 stand for the number of dollars you earned, and ⁻5 stand for the number of dollars you spent, then the sum is $3 + (^-5) = ^-2$. The fact that the sum is a negative integer shows that after you pay out all your earnings you are still in debt.

When you add the integers that stand for money received and money spent, a positive sum means you have money left, and shows how much is left; a negative sum means that you are in debt, and shows how much you owe.

The drawings below show four people's records of what they earned and what they spent during a week. Add the integers to find out how much money each has left, or how much money he owes, after accounts are settled.

Tommy		Jane	
Dollars Earned	Dollars Spent	Dollars Earned	Dollars Spent
5	−5	4	−5

Ruth		Omar	
Dollars Earned	Dollars Spent	Dollars Earned	Dollars Spent
4	−2	3	−6

A Slide Rule for Addition

Two copies of the number line can be used to make a slide rule for the addition of integers. To make a slide rule, use a sheet of thin, stiff cardboard, that can be creased without breaking, and follow these directions:

Cut a strip of cardboard that is twelve inches long and two inches wide. Draw a straight line down the middle of the strip, so that it is one inch from each of the long edges. Put a ruler on the strip so that the edge of the ruler is on the line. While you press the ruler down firmly with one hand, use your other hand to bend up against the edge of the ruler the half of the strip that is not under the ruler. Remove the ruler and bring the two halves of the strip together so that their edges meet. Rub the end of the ruler along the crease to flatten it. Now, make a series of short lines along one of the cut edges of the folded strip, to divide it into half-inch spaces. Write 0 under the line that is at the center of the strip. Then put the positive integers on the right, and the negative integers on the left, as shown. The edge will then show the part of the number line that extends from ⁻11 to 11.

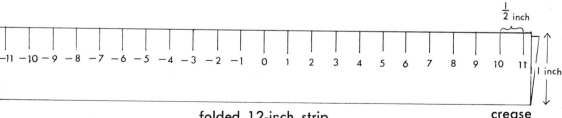

folded 12-inch strip

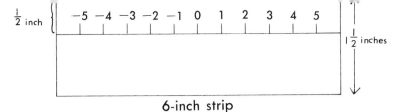

6-inch strip

Next cut a strip of cardboard that is six inches long and 1½ inches wide. Draw a straight line one half inch from one edge. Along this line make a series of short lines that divide it into half-inch spaces, and label them to show the part of the number line that extends from ⁻5 to 5, as shown in the drawing. Now insert this strip in the fold of the longer strip, so that the part with the numbers on it is on top. U. S. 1719344

To add two integers, put the 0 of the short strip over the point on the long strip where you see the first of the two integers that are to be added. Now locate the second integer on the short strip. The answer is right under it, on the long strip. The drawing on this page shows the setting for adding 3 + (⁻5).

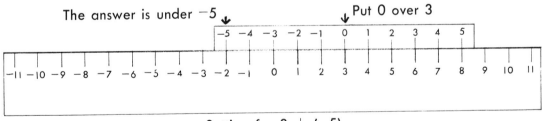

Setting for 3 + (⁻5)

Use your slide rule to do these addition exercises:

1 + 3	2 + (⁻4)	(⁻3) + (⁻4)
5 + 0	3 + (⁻3)	(⁻4) + 1
0 + (⁻3)	(⁻2) + (⁻1)	(⁻5) + 4

An Adding Machine

You can make a simple adding machine that adds integers and shows you the answer through a window. To make the machine, use thin, stiff cardboard, and follow these directions.

Cut a strip of cardboard twelve inches long and five inches wide. Draw a line down the middle so that it is 2½ inches from each long edge. Fold the strip on this line, and press the crease flat. Draw a line one half inch from the cut edge, and draw another line one inch from that edge. Between these two lines draw two half-inch squares separated by a space of 1½ inches, as shown in the drawing on page 37. Then cut out the squares. Press the front half of the folded strip flat against the back half, and trace the outlines of the cut-out squares on the back half. Then cut out the squares on the back half. Lay the strip, folded and flat, on the table, with the crease at the left. To the left of the upper square, write 4 +; to the left of the lower square write *Answer window*. Now turn the folded strip upside down, but keep the crease on the left. On the side that is face up now, write ‾4 + to the left of the lower square, and write *answer window* to the left of the upper square. The two sides of the folded strip are your *window cards* for adding any integer to 4 and to ‾4).

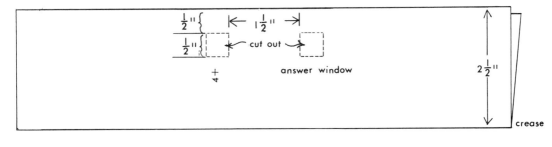

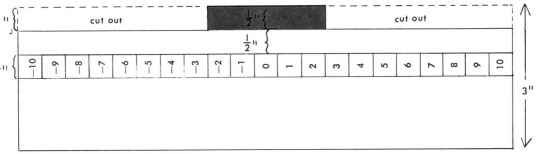

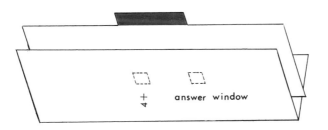

You also need a *slide card* to insert between these two window cards. To make the slide card, cut a strip of cardboard that is twelve inches long and three inches wide. Along one edge, draw three lines to make three strips that are one-half inch wide. In the middle of the strip nearest the edge, draw the outline of a tab, and color it red as shown in the drawing. Cut away the rest of the strip on each side of the tab. Be careful not to cut the tab off. Divide the innermost strip into half-inch squares. Write 0 in the middle square. Write the positive integers in order in the squares below the zero square. Write the negative integers in order in the squares above the zero square. Now put the slide card between the pair of window cards. Use the tab for moving the slide card up or down. If the 4 + side is face up, you are adding to 4 the number that shows through the upper window. If the ⁻4 + side is face up, you are adding to ⁻4 the number that shows through the lower window. In each case the answer is shown through the answer window.

To make window cards for 2 + and ⁻2 +, make the space between the windows one half inch long. For 3 + and ⁻3 +, make the space between the windows one inch long. For 5 + and ⁻5 +, make the space two inches long. How long should you make the space for 6 + and ⁻6 +? For 7 + and ⁻7 +? For 8 + and ⁻8 +?

A Rule for Addition

There is a way of adding integers in your head by following two simple rules. Let us find the rules.

On page 25 we introduced the idea of the *absolute value* of an integer. It is the number of units in the distance from 0 to that integer on the number line. If we picture an integer by means of checkers, the absolute value of the integer is the number of checkers that are used. Now we shall use this idea to develop the rules for addition.

Rule 1. *Suppose we are adding two numbers that are either both positive or both negative.* If we picture them as sets of checkers, we use the same color checkers for both sets. When we unite the sets of checkers, we are adding the number of checkers in one set to the number of checkers in the other set. That is, *we are adding the absolute values of the integers.* The color of the united set is the same as the color that we started with. So *if the two integers we add are positive, their sum is positive. If the two integers we add are negative, their sum is negative.* The sentences above in italics give rule 1 for addition.

For example, to add the positive integers 2 and 3, simply add the absolute values 2 and 3, and the sum is 5, a positive integer. To add the negative integers, ⁻2 and ⁻3, add the absolute values 2 and 3, and the sum is ⁻5, a negative integer.

Rule 2. *If one integer is positive and one is negative,* we picture them as sets of checkers of different colors. We unite the two sets of checkers, and then take away matched pairs, one red and one black, until all the checkers left are of one color. The color of those that are

-2 ⚫⚫
-3 ⚫⚫⚫

Add the absolute values:
2 + 3 = 5
So, (⁻2) + (⁻3) = ⁻5

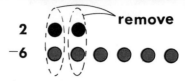

2 ⚫⚫
⁻6 ⚫⚫⚫⚫⚫⚫

Subtract the absolute values:
6 − 2 = 4
So, 2 + (⁻6) = ⁻4

left is the color of the set that had the larger number of checkers. That it, it is the color of the set for the integer that has the larger absolute value. So, *if the integer with the larger absolute value is positive, the sum of the integers is positive. If the integer with the larger absolute value is negative, the sum of the integers is negative.* The number of checkers that are left is the number that remains after you take away from the set with the larger number of checkers (larger absolute value) as many checkers as there are in the set with the smaller number of checkers (smaller absolute value). So *you subtract the smaller absolute value from the larger absolute value to get the absolute value of the sum.* The sentences above in italics give rule 2 for addition.

For example, to add the positive integer 2 and the negative integer ⁻6, subtract the smaller absolute value 2 from the larger absolute value 6 to get a difference of 4. Then the sum of the integers is ⁻4, since the negative number ⁻6 has the larger absolute value.

Use rules 1 and 2 to add these integers:

8 + (⁻3)	(⁻2) + (⁻2)	4 + (⁻6)
(⁻2) + 5	(⁻7) + 1	(⁻5) + 5

Subtraction of Integers

Subtraction is like addition done backwards. For example, to subtract 2 from 7 means to find the integer you must add to 2 to get 7 as the sum. This is easy to do with checkers. First put out two black checkers to stand for 2. Then put out more black checkers until there are seven black checkers altogether. Since you put out 5 more black checkers to get 7, we say that 2 subtracted from 7 leaves 5. We write $7 - 2 = 5$.

To subtract $^-3$ from $^-5$ means to find the integer you must add to $^-3$ to get $^-5$. First put out three red checkers to stand for $^-3$. Then put out more red checkers until there are five red checkers altogether. Since you put out 2 more red checkers to get $^-5$, we say that $^-3$ subtracted from $^-5$ leaves $^-2$. We write $(^-5) - (^-3) = (^-2)$. Notice that in this statement we use the minus sign in two different ways. The second minus sign means *subtract*. The other minus signs are parts of the integers and show that these integers are negative.

To subtract $^-2$ from 3 means to find the integer you must add to $^-2$ to get 3. First put out two red checkers to stand for $^-2$. Since you aim to get 3, which is pictured by black checkers, you put out black checkers next. The first two black checkers put out will not stay, because they will be removed with the two red checkers

Start with 2
black checkers

Put out more black
checkers until there
are 7 blacks

2 + 5 = 7

7 − 2 = 5

Start with 3
red checkers

Put out more red
checkers until there
are 5 reds

(−3) + (−2) = −5

(−5) − (−3) = (−2)

to be removed

Start with 2
red checkers

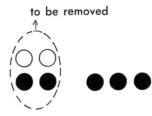

Put out more black
checkers until there
are 3 that stay

(−2) + 5 = 3

3 − (−2) = 5

→ to be removed

Start with 3
black checkers

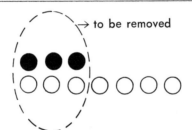

Put out red
checkers until there
are 4 that stay

3 + (−7) = (−4)

(−4) − 3 = (−7)

that they cancel. The next three black checkers put out
will stay, and give you the three you want. Since you
put out five black checkers in all to get 3, we say that ⁻2
subtracted from 3 leaves 5. We write 3 − (⁻2) = 5.

To subtract 3 from ⁻4 means to find the integer you
must add to 3 to get ⁻4. First put out three black
checkers to stand for 3. Since you aim to get ⁻4, which

is pictured by red checkers, you put out red checkers next. The first three red checkers you put out will not stay, because they will be removed with the three black checkers that they cancel. The next four red checkers put out will stay, and give you the four you want. Since you put out seven red checkers in all to get $^-4$, we say that 3 subtracted from $^-4$ leaves $^-7$. We write $(^-4) - 3 = (^-7)$.

Use checkers to find the answers to these questions:

How many black checkers added to four black checkers will give you six black checkers altogether?

How many red checkers added to two red checkers will give you three red checkers altogether?

How many black checkers added to two red checkers will give you four black checkers left, after the checkers that cancel are removed?

How many red checkers added to three black checkers will give you four red checkers left, after the checkers that cancel are removed?

What color checkers should you add to six black checkers to give you two black checkers altogether? How many of them should you add?

Use checkers to do these subtraction exercises:

$$1 - (^-2) \qquad (^-1) - (^-4) \qquad 2 - 8$$
$$(^-2) - (^-2) \qquad 4 - 0 \qquad 5 - 2$$
$$0 - 3 \qquad 2 - (^-6) \qquad 3 - (^-3)$$

Subtraction on the Number Line

There is an easy way to do subtraction of integers on the number line.

Suppose you want to subtract $^-2$ from 5. Then you must find the integer that you should add to $^-2$ to get 5. First locate $^-2$ on the number line, and move from $^-2$ to 5. Then you are doing one-step addition. The number you are adding to $^-2$ is shown by the number of units you moved and the direction in which you moved. The diagram shows that you moved 7 units to the right. So the number you added to $^-2$ to get 5 is 7. That is, $5 - (^-2) = 7$.

$5 - (^-2) = 7$ $^-5$ $^-4$ $^-3$ $^-2$ $^-1$ 0 1 2 3 4 5 6 7

Suppose you want to subtract 8 from 6. First locate 8 on the number line, and move from 8 to 6. You have to move 2 units to the left. So $6 - 8 = ^-2$.

$6 - 8 = ^-2$ 0 1 2 3 4 5 6 7 8 9

Suppose you want to subtract $^-3$ from $^-2$. First locate $^-3$ on the number line, and then move from $^-3$ to $^-2$. You have to move 1 unit to the right. So $(^-2) - (^-3) = 1$.

$^-8$ $^-7$ $^-6$ $^-5$ $^-4$ $^-3$ $^-2$ $^-1$ 0

To subtract on the number line, move from the number you are subtracting to the number you are subtracting it from. The number of units in your motion and the direction of the motion picture the answer.

What subtraction exercise is pictured by each drawing below?

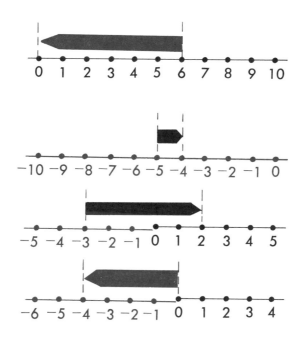

Do each of these subtraction exercises on the number line:

$3 - 8$	$2 - (^-4)$	$4 - 4$
$(^-5) - (^-5)$	$0 - (^-6)$	$6 - 3$
$0 - 3$	$(^-2) - (^-5)$	$(^-5) - (^-2)$

A Rule for Subtraction

There is a simple rule by which you can change every subtraction exercise into an addition exercise that produces the correct answer.

To discover this rule, we shall first do some subtractions on the number line in a new way. Instead of moving in one step from the number that is being subtracted to the number that we are subtracting it from, we shall move in two steps, moving to zero first.

For example, to subtract 3 from 2, first move from 3 to 0, and then from 0 to 2. These two motions, shown as arrows in the drawing, picture the integers $^-3$ and 2. The sum of these two integers, $(^-3) + 2$ is the integer pictured by the arrow that is left after we remove the two black arrow units from 0 to 2 and the two red arrow units from 2 to 0 that cancel each other. This arrow that is left, running from 3 to 2, is the answer to the subtraction example. So we have obtained the answer to the subtraction example $2 - 3$ by doing the

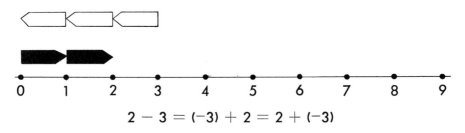

$$2 - 3 = (^-3) + 2 = 2 + (^-3)$$

46

addition $(^-3) + 2$. But $(^-3) + 2 = 2 + (^-3)$. Notice that $^-3$ is the opposite of 3. So we have found that subtracting 3 is like adding its opposite.

To subtract $^-4$ from 3, first move from $^-4$ to 0, and then from 0 to 3. The diagram shows that subtracting $^-4$ is like adding its opposite, 4.

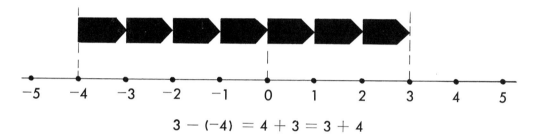

$$3 - (^-4) = 4 + 3 = 3 + 4$$

In general, to subtract any integer, simply add its opposite.

Do these subtractions on the number line by moving in two steps, moving to zero first.

Subtract 4 from $^-2$.　　Subtract $^-1$ from 5.

Do these subtractions by following the rule that to subtract a number you simply add its opposite.

$2 - 7 = 2 + (^-7) = $ _____

$3 - (^-5) = 3 + $ _____ $ = $ _____

$(^-2) - (^-8) = ^-2 + $ _____ $ = $ _____

$0 - (^-4) = 0 + $ _____ $ = $ _____

$2 - 0 = 2 + $ _____ $ = $ _____

$(^-3) - (2) = ^-3 + $ _____ $ = $ _____

About the Author

The *Reason Why* Books were initiated by Irving and Ruth Adler, who worked together to write the first thirty books in the series. This book, the thirty-fourth, is the third one written by Irving Adler since his wife, Ruth, died in 1968.

Irving and Ruth Adler wrote jointly or separately eighty books about science and mathematics. Dr. Adler has been an instructor in mathematics at Columbia University and at Bennington College, and was formerly head of the mathematics department of a New York City high school. Ruth Adler taught mathematics, science, and art in schools in the New York area, and later also taught at Bennington. In addition to working with her husband writing *Reason Why* books, she drew the illustrations for most of them as well as for many other books written by him.

Books by Irving Adler alone and books by him in collaboration with Ruth Adler have been printed in 117 different foreign editions, in fifteen languages and in ten reprint editions.